Language for maths

y5

smaller

Altogether

Greater

combined

Share

each

less

increase

Help pupils untangle SATs reasoning questions

Contents

How to use this book

This book is divided into five sections: Number – number and place value; Number – addition and subtraction; Number – multiplication and division; Number – fractions; Explain why.

Each section consists of: instructions for a number of collaborative activities; a range of sample questions (based on papers 2 and 3 from the 2016 and 2017 SATs) plus worked support examples to guide pupils, and answers for all questions. Accompanying printable resources and answers for the collaborative activities are available via the free online download.

Pupils should undertake the collaborative activities first in order to rehearse the language needed when answering the supported questions.

Collaborative activities

Collaborative activities are learning activities that are designed for the purpose of helping pupils to understand and use vocabulary and sentence structures which are relevant and important to the subject. They are also intended to enable pupils to better understand the concepts and processes involved in solving problems. In this way, the language and content of the subject matter are integrated.

There are a number of different strategies for creating opportunities for pupils to understand and use language in the context of mathematics. Each of the strategies creates the need for pupils to ask each other questions and make statements about the topic.

Track games

Track games are designed so that pupils have to ask and answer a range of questions within a game-like format. At the same time, pupils are supported in forming both the questions and the answer statements. Pupils may also be motivated by the idea of 'winning' the game.

Card chains, card sequences and domino chains

These activities are intended to enable pupils to link previous information to new information and to get pupils to calculate on the basis of the new information. Card chains can also be used to encourage pupils to think about sequence. For example, they encourage pupils to think and talk about the order of numbers in terms of lowest to highest.

Card sorting

These activities require pupils to sort cards into categories. For example, they may require pupils to sort the cards into the appropriate columns on a grid or the appropriate section on a sorting table. They encourage pupils to differentiate between cards which contain similar but different information.

Card matching

These activities support pupils in arranging a set of cards on a grid so that the information on the cards matches across a row. For example, they may involve matching a question card with an answer card or a calculation card with an explanation card.

Clue sheets

Clue sheets require pupils to read and share clues which help them to complete a table or diagram. The clue sheets have different clues on them so that pupils have to cross-reference clues in order to complete the task. A task for four pupils may involve four different clue sheets so that each pupil has to contribute.

Information gap / barrier games

Information gap games involve pupils being given some information, but they will also be missing some important information. So Pupil A may have half the information needed to complete a task while Pupil B has the other half. This creates the need for pupils to ask and answer each other's questions in order to complete the task. For example, Pupil A may have half of the bars on a bar chart and will have to ask Pupil B questions so that he / she can fill in the missing bars on the chart accurately and, at the same time, Pupil B will have to ask Pupil A questions to obtain the missing information that he / she needs.

Four in a line games

These activities are useful for practising vocabulary and sentence structures and encouraging flexibility of thinking. There are a number of different rectangles on a game board where each supplied card could be placed. Rather than simply choosing the first possible rectangle, pupils need to scan the board to see if they can find a rectangle which gives a greater advantage in terms of connecting four of their cards and thereby winning the game. The game-like nature of the activity encourages natural repetition of vocabulary and sentence structures and also often means that pupils want to 'play again'.

Substitution tables

Substitution tables are useful for supporting pupils' abilities to form relevant questions and statements. Pupils generate a range of questions for a partner to answer. Pupils could also work together to form questions for another group or pair. This requires them to be sure that they know the correct answers to their questions. Pupils can also devise statements and give them to other pupils to check. A variation on this is to ask pupils to devise a set of statements, some of which are true and some of which are false. Another group of pupils can then check the statements and change the false statements to make them true. Substitution tables can be used to encompass a wide range of topics to familiarise pupils with typical vocabulary and sentence structures relating to these topics.

Collaborative guided talk

These activities aim to engage pupils in a small-group dialogue which is initially led by a teacher or teaching assistant. The adult asks questions and provides prompts to shape the thinking and language around a particular problem. This also involves modelling the appropriate language and re-casting what pupils say. As the activity proceeds, the intention is for pupils to 'take over' the questioning and engage in a greater proportion of the dialogue.

Number – number and place value

Collaborative activities

Four in a line: More than or less than

An activity for two pupils or four pupils in two teams of two. *See download for accompanying printable resources.*

Content: 100 / 1000 / 10 000 more than or less than a six-digit number

Key language

Three hundred and fifty thousand two hundred and twenty-one **is 1000 more than** three hundred and forty-nine thousand two hundred and twenty-one.

Three hundred and forty-nine thousand two hundred and twenty-one **is 100 less than** three hundred and forty-nine thousand three hundred and twenty-one.

Instructions

1. Cut the cards into individual cards.
2. One pupil or team has the yellow cards and the other pupil or team has the green cards.
3. Place the cards face down in two piles.
4. The yellow team takes the top card from their pile and places it on a possible box on the board. When they put the card down, they must say a correct sentence (e.g. "Three hundred and forty-nine thousand two hundred and twenty-one is one hundred less than three hundred and forty-nine thousand three hundred and twenty-one.").
5. The green team then places their card and says a sentence.
6. The aim of the game is to get four cards of your colour in a line – across, down or diagonally.
7. The winner is the first team to get four of their colour in a line.

Clue sheets: Counting across zero

An activity for two pupils. *See download for accompanying printable resources.*

Content: Directed (negative) numbers: counting forwards and backwards across zero

Key language

The temperature in Oslo was **minus 7°C**.

The temperature in Naples was **2 degrees higher than** in Paris.

The temperature in Belgrade was **4 degrees lower than** in Rome.

Context

Pupils complete a chart to show the temperature in 17 cities in Europe at 10 a.m. on one day in January.

Instructions

1. Pupil A has clue sheet A and Pupil B has clue sheet B. The pupils have a copy of the board and a set of city cards between them.

2 Pupils take turns to read a clue from their clue sheet and decide together where a city card should be placed on the board.

3 Encourage pupils to justify their decision (e.g. "If the temperature in Naples was 2° higher than in Paris, then the temperature in Naples must be 6°C.").

Information gap / barrier game: Saying and ordering large numbers

An activity for two pupils. *See download for accompanying printable resources.*

Content: Saying large numbers in words and ordering large numbers

Key language

What was the attendance at the Arsenal match? The attendance at the Arsenal match was **fifty-nine thousand, nine hundred and fifty-seven people**.

The attendance at Arsenal was **higher** than at Everton.

The **highest** attendance was at Manchester United.

The **lowest** attendance was at Southampton.

Context

Pupils complete a table showing the number of people (the attendance) at different football matches.

Instructions

1 Pupil A has sheet A and Pupil B has sheet B.

2 Pupil A has information on their sheet which Pupil B does not have and Pupil B has information which Pupil A does not have.

3 Pupils ask each other questions to fill in the missing information in the 'Attendance in numbers' column. They read the attendance figures aloud and record them on their sheet.

4 They then use this information to write the numbers in words in the 'Attendance in words' column.

5 Finally, pupils work together to order the attendance from highest to lowest and complete the final column.

Domino chains: More than and less than

An activity for two pupils. *See download for accompanying printable resources.*

Content: 100, 1000, 10 000 and 100 000 more than and less than

Key language

239 673 is **7000 more than** 232 673.

322 673 is **4000 less than** 326 673.

Instructions

1 Cut out the dominoes.

2 Share out the dominoes between the pupils.

3 Together, pupils try to make as long a chain of inequalities as possible and to join up all the dominoes. The statement on the right-hand side of the domino must be joined to the correct number on the left-hand side of the next domino in the chain.

4 Encourage pupils to verbalise their chain (e.g. "239 673 is 7000 more than 232 673.").

Note: The statement on the right-hand side of the last domino must match the number on the left-hand side of the first domino. In this way, the chain forms a complete loop.

Clue sheets: Roman numerals

An activity for two pupils. *See download for accompanying printable resources.*

Content: Roman numerals and dates

Key language

When was Diana born? What date is that **in Roman numerals**?

What does L / C / M etc. **mean** in numbers?

Instructions

1 Pupil A has clue sheet A and Pupil B has clue sheet B. Each pupil has a copy of the recording sheet.

2 Pupils take turns to read a clue from their clue sheet and decide together where the name of the person should go on the sheet so that the name goes with the correct Roman numeral date.

3 They complete the table by writing the years in numbers below the Roman numerals.

Note: The task can be made more demanding by making some of the clues comparative rather than definitive. For example, "Maria was born three years before Alex" or "Juliana is 12 years older than David". This increases the range of language involved.

Clue sheets: Reading and rounding large numbers

An activity for two pupils. *See download for accompanying printable resources.*

Content: Reading large numbers in words and rounding large numbers to the nearest 100, 1000, 10 000 and 100 000

Key language

The population of Sheffield is **five hundred and thirty-five thousand, seven hundred and eighty-two**.

The population of Sheffield is five hundred and thirty-six thousand **rounded to the nearest thousand**.

Context

Pupils complete a table showing the population of some cities in the UK. The populations are for city boundaries and not metropolitan areas.

Instructions

1 Pupil A has clue sheet A and Pupil B has clue sheet B. Each pupil has a copy of the recording sheet.

2 Pupils take turns to read a clue from their clue sheet and decide together where to record the information in the population column on their recording sheet.

3 They then work out the populations of the cities rounded to the nearest 100 000, 10 000, 1000 and 100.

4 Encourage pupils to verbalise their decisions (e.g. "The population of Sheffield is five hundred and thirty-six thousand rounded to the nearest thousand.").

Questions with support

1 Write these Roman numerals in figures.

CCXV	DCCCXX	CM	LIX

>  "Write the Roman numeral in figures" means write the Roman numeral in numbers, for example, 2017.
>
> Remember that **M** is 1000, **D** is 500, **C** is 100, **L** is 50, **X** is 10, **V** is 5 and **I** is 1. Therefore, **MMXVI** is 1000 + 1000 + 10 + 5 + 1 = **2016**.
>
> You also need to know that, if **a letter for a smaller number** is **before a letter for a larger number**, you have to **subtract** the smaller number from the larger number. Therefore:
>
> | IV is 5 − 1 = 4 | IX is 10 − 1 = 9 | XC is 100 − 10 = 90 |
> |---|---|---|
>
> | XL is 50 − 10 = 40 | CD is 500 − 100 = 400 | CM is 1000 − 100 = 900 |
> |---|---|---|

2 At the end of these television programmes, the year that the programme was made is given in Roman numerals. Write the years in figures under the programmes.

Animal Adventure MMXI	The Lost City MMIX	Music School MMVI

>  "Write **the year in figures**" means write the year **in numbers**, for example, 2017.

3 **a.** What number is one thousand more than 231 540?

┌─────────────┐
│   │
│   │
└─────────────┘

b. What number is one hundred less than 154 378?

┌─────────────┐
│   │
│   │
└─────────────┘

c. What number is one thousand less than 110 000?

┌─────────────┐
│   │
│   │
└─────────────┘

d. What number is one hundred more than 89 900?

┌─────────────┐
│   │
│   │
└─────────────┘

 Remember that **more than** in this kind of question means **add**. So **100 more than 216** is 216 plus 100. Therefore **100 more than** 216 is 316.

1000 more than 216 is 216 plus 1000. Therefore **1000 more than** 216 is 1216.

Less than in this kind of question means subtract. So, **100 less than** 3452 is 3452 minus 100. Therefore **100 less than** 3452 is 3352.

1000 less than 3452 is 3452 minus 1000. Therefore **1000 less than** 3452 is 2452.

(4) Complete the table.

Number	10 times greater	100 times greater
243		
3107		
2876		

> **10 times greater** than a number means you need to **multiply by 10**. So, **10 times greater than 5** equals 10 multiplied by 5. Therefore **10 times greater than 5** equals 50.
>
> **100 times greater than 5** equals 100 multiplied by 5. Therefore **100 times greater than 5** equals 500.

(5) In April, Zahir had 203 hits on his internet blog.

a. In May, the hits were 100 times greater than in April. How many hits did Zahir have in May?

b. In August, the hits were 1000 times greater than the number of hits in April. How many hits did Zahir have in August?

(6) Write these numbers in order. Start with the smallest number.

465 678   444 563   399 678   465 599   462 214

Smallest   **Largest**

> When ordering numbers, it can be useful to write the numbers in **columns**. It may help to write the numbers in a table like this, which shows the **place value** of each **digit**.
>
> | Place value | | | | | |
> |---|---|---|---|---|---|
> | Hundred thousands | Ten thousands | Thousands | Hundreds | Tens | Ones |
> | | | | | | |
> | | | | | | |
> | | | | | | |
>
> Look at the **hundred thousands** first. Is there a number with **fewer hundred thousands** than the other numbers? This is the **smallest** number. Then look at the **ten thousands**. Is there a number with **fewer ten thousands** than the other numbers? This is the next smallest number.
>
> Then compare the other three numbers by looking at the thousands, hundreds, tens and ones.

7 Look at these six numbers.

456 768   665 432   563 474   948 217   217 321   735 234

a. Which number has three hundreds?

b. Which number has three thousands?

c. Which number has four ten thousands?

d. Which number has six hundred thousands?

8 These are the prices of five cars.

Car A	Car B	Car C	Car D	Car E
£32 999	£33 000	£34 500	£30 800	£29 950

Put them in order of price. Start with the lowest price.

£	£	£	£	£

Lowest   **Highest**

> The **lowest** price is the **smallest** number. The **highest** price is the **largest** number.

9 359 269 people live in the city of Coventry.

How many people live in Coventry rounded to the nearest 10, 100, 1000 and 10 000?

Rounded to the nearest 10

Rounded to the nearest 100

Rounded to the nearest 1000

Rounded to the nearest 10 000

70	71	72	73	74	75	76	77	78	79	80
Rounded to the nearest 10 are all 70					Rounded to the nearest 10 are all 80					

400	410	420	430	440	450	460	470	480	490	500
Rounded to the nearest 100 are all 400					Rounded to the nearest 100 are all 500					

3000	3100	3200	3300	3400	3500	3600	3700	3800	3900	4000
Rounded to the nearest 1000 are all 3000					Rounded to the nearest 1000 are all 4000					

60 000	61 000	62 000	63 000	64 000	65 000	66 000	67 000	68 000	69 000	70 000
Rounded to the nearest 10 000 are all 60 000					Rounded to the nearest 10 000 are all 70 000					

Therefore 75 rounded to the nearest 10 is 80.

Therefore 475 rounded to the nearest 100 is 500.

Therefore 3475 rounded to the nearest 1000 is 3000.

Therefore 63 475 rounded to the nearest 10 000 is 60 000.

63 475 rounded to the nearest 10 is 63 480.

63 475 rounded to the nearest 100 is 63 500.

63 475 rounded to the nearest 1000 is 63 000.

63 475 rounded to the nearest 10 000 is 60 000.

10 The table below shows the temperature in °C in three cities on Monday and Friday in one week.

	London	Warsaw	Prague
Monday	2°C	−4°C	−6°C
Friday	−1°C	1°C	−2°C

a. How much higher was the temperature on Monday in London than in Warsaw?

b. How much lower was the temperature on Monday in Prague than in London?

c. By how much did the temperature rise in Warsaw from Monday to Friday?

d. By how much did the temperature fall in London from Monday to Friday?

In the example below, the temperature **rose** from minus 4 degrees C to 1 degree C. The **difference** in temperature was 5 degrees. It was 5 degrees **higher**.

−7°C	−6°C	−5°C	−4°C	−3°C	−2°C	−1°C	0°C	1°C	2°C	3°C
				1	2	3	4	5		

In this example, the temperature **fell** from 2 degrees C to minus 5 degrees C. The **difference** in temperature was 7 degrees. It was 7 degrees **lower**.

−7°C	−6°C	−5°C	−4°C	−3°C	−2°C	−1°C	0°C	1°C	2°C	3°C
		7	6	5	4	3	2	1		

Answers

1. CCXV = 215   DCCCXX = 820   CM = 900   LIX = 59

2.

Animal Adventure	The Lost City	Music School
MMXI	MMIX	MMVI
2011	2009	2006

3. **a.** 232 540   **b.** 154 278   **c.** 109 000   **d.** 90 000

4.

Number	10 times greater	100 times greater
243	2430	24 300
3107	31 070	310 700
2876	28 760	287 600

5. **a.** 20 300   **b.** 203 000

6.

399 678	444 563	462 214	465 599	465 678

Smallest   **Largest**

7. **a.** 217 321   **b.** 563 474   **c.** 948 217   **d.** 665 432

8.

£29 950	£30 800	£32 999	£33 000	£34 500

Lowest   **Highest**

9.

359 270	359 300	359 000
Rounded to the nearest 10	Rounded to the nearest 100	Rounded to the nearest 1000

360 000
Rounded to the nearest 10 000

10. **a.** 6 degrees   **b.** 8 degrees   **c.** 5 degrees   **d.** 3 degrees

Number – addition and subtraction

Collaborative activities

Card sorting: Addition and subtraction combinations

An activity for two, three or four pupils. *See download for accompanying printable resources.*

Content: Addition and subtraction and exploring combinations of numbers which when added or subtracted produce the same answer

Key language

Six thousand plus five thousand equals eleven thousand.

Eight thousand plus three thousand also equals eleven thousand.

Eleven thousand minus nine thousand equals two thousand.

Ten thousand minus eight thousand also equals two thousand.

Context

There are two similar activities. One involves addition and the other subtraction. Pupils should do the addition activity first and then the subtraction activity.

Instructions

1. Cut the cards into individual cards.
2. Share out the cards among the pupils.
3. Pupils work together to place the cards on the grids in order to make correct calculations.
4. The aim is to place all the cards on the grids. Because there are a number of possibilities for making correct calculations, it is necessary to find the right combinations in order to be able to place all the cards. In this way, pupils have to explore a number of different combinations to achieve the same answer.

Information gap / barrier game: Unknown quantities

An activity for two pupils. *See download for accompanying printable resources.*

Content: Finding the value of unknown numbers in column addition and subtraction

Key language

How many thousands / hundreds / tens / ones are there in the first number?

If there are 3 ones in the first number **and 2 ones** in the answer, **how many ones must there be** in the second number?

There must be 9 ones in the second number **because** 3 plus 9 equals 12.

There must be 7 ones in the second number **because** 14 minus 7 equals 7.

Instructions

1. Pupil A has sheet A and Pupil B has sheet B. Pupil A has sinformation on their sheet which Pupil B does not have and Pupil B has information which Pupil A does not have.
2. Pupils ask each other questions to fill in their missing information (e.g. "In question 1, how many tens are there in the first number?").
3. When they have exchanged all the information that they have, they work together to find the missing numbers in the calculations.

Information gap / barrier game and substitution tables: Differences in distances

An activity for two pupils. *See download for accompanying printable resources.*

Content: Calculating differences in distances (four-digit values)

Key language

How far is it from London **to** Islamabad? It's six thousand and forty-one kilometres **from** London **to** Islamabad.

How much further is it from London **to** Athens than **from** London **to** Bucharest? It's three hundred and two kilometres further.

How much nearer is it from London to Cairo than **from** London **to** Nairobi? It's three thousand three hundred and fourteen kilometres nearer.

Instructions

1. Pupil A has grid A and Pupil B has grid B.

2. Pupil A has information on their grid which Pupil B does not have and Pupil B has information which Pupil A does not have.

3. Pupils ask each other questions to fill in their missing information.

4. When the pupils have completed the grids, they use the substitution tables and the information in the grids to create 10 questions and 10 true statements.

Information gap / barrier game: Rounding numbers

An activity for two pupils. *See download for accompanying printable resources.*

Content: Using rounding to estimate calculations

Key language

2154 **rounded to the nearest 100** is 2200 and 4149 **rounded to the nearest 100** is 4100 so 2154 plus 4149 is **approximately** 6300.

6459 **rounded to the nearest 100** is 6500 and 3236 **rounded to the nearest 100** is 3200 so 6459 minus 3236 is **approximately** 3300.

Instructions

1. Pupil A has grid A and Pupil B has grid B.

2. Pupil A has information on their grid which Pupil B does not have and Pupil B has information which Pupil A does not have.

3. Pupils ask each other questions to fill in their missing information.

4. They then have to agree the answers to the missing information that neither of them has, using the information that they do have.

Clue sheets: Changes and comparison

An activity for two pupils. *See download for accompanying printable resources.*

Content: Addition and subtraction relating to change and comparison word problems

Key language

At the beginning of April, a shop had 71 tennis rackets. In April, it had a delivery of 64 netballs.

During April, it sold 236 shirts. At the end of April, it had 44 more shirts than at the beginning of April.

The shop must have had _ netballs at the beginning / end of April.

The shop must have had a delivery of / sold _ netballs in / during April.

Context

The Super Sports Shop had six kinds of sports equipment in stock at the beginning of April. It had a delivery in April and sold some items during April. At the end of April, it had some items left. Pupils use the information supplied to calculate the missing information.

Instructions

1. Pupil A has clue sheet A and Pupil B has clue sheet B. Each pupil has a copy of the recording sheet.
2. Pupils take turns to read a clue from their clue sheet and decide together where to write the information on the recording sheet so that the number is placed in the correct row and column.
3. When they have recorded all the information from the clues, the pupils work together to calculate the missing information.

Note: The clues model the typical language of two-step addition and subtraction word problems. The table also illustrates the semantic structure of two-step additive change problems. Problems of this kind can ask for the answer to either the beginning or first change or second change or the result. Using a table can help pupils become more aware of the semantic structure of these problems.

Information gap / barrier game and substitution tables: Ticket prices

An activity for two pupils. *See download for accompanying printable resources.*

Content: Asking and answering questions about the cost of train tickets

Key language

How much does it cost to buy an **adult / child single / return ticket** to Leeds? It costs £ _____ .

How much does it cost to buy **two adult return tickets and three child return tickets** to Liverpool?

Context

Pupils complete grids showing the cost of single and return train fares for adults and children from Manchester to three different cities.

Instructions

1. Pupil A has grid A and Pupil B has grid B. Pupil A has information on their grid which Pupil B does not have and Pupil B has information which Pupil A does not have.
2. Pupils ask each other questions to fill in their missing information.
3. When the pupils have completed the grids, they use the substitution tables and the information in the grids to create 10 questions and 10 true statements.

Number – addition and subtraction

Questions with support

1 Mrs Jones had £35. She spent £8 at the supermarket. Then she spent some more money at the clothes shop. After that she had £7 left. How much did she spend in the clothes shop?

```
┌─────────────────┐
│                 │
│                 │
└─────────────────┘
```

> ! What numbers do you know from the question that you can write in the boxes below? What numbers are unknown?
>
> Can you calculate the unknown numbers? Do you have to add or subtract to calculate the unknown numbers?
>
> | In the first place, she had | Then she spent | After that she had | Then she spent | After that she had |
> |---|---|---|---|---|
> | | | | | |

2 Pavel had some football cards. He lost 16 of them, but then his friend gave him 24 more cards. Now he has 45 cards. How many cards did he have in the first place?

```
┌─────────────────┐
│                 │
│                 │
└─────────────────┘
```

> ! What numbers do you know from the question that you can write in the boxes below? What numbers are unknown?
>
> Can you calculate the unknown numbers? Do you have to add or subtract to calculate the missing numbers?
>
> | In the first place, he had | Then | After that he had | Then | Now |
> |---|---|---|---|---|
> | | | | | |

3 Shahida had some stickers. Azam had 39 stickers. Shahida had 12 more stickers than Azam. Then Shahida bought 24 more stickers. How many stickers does she have now?

> ! What numbers do you know from the question that you can write in the boxes below? What numbers are unknown?
>
> Can you calculate the missing numbers? Do you have to add or subtract to calculate the missing numbers?
>
> | Who had more stickers? How many? | | How many more stickers? | Then | Now |
> |---|---|---|---|---|
> | Who had fewer stickers? How many? | | | | |

4 Agata had 36 books. She had 15 more books than Maria. Then Maria gave 6 of her books to her friend. How many books does Maria have now?

> ! What numbers do you know from the question that you can write in the boxes below? What numbers are unknown?
>
> Can you calculate the unknown numbers? Do you have to add or subtract to calculate the unknown numbers?
>
> | Who had more? How many? | | How many more? | Then | Now |
> |---|---|---|---|---|
> | Who had fewer? How many? | | | | |

5 There were 54 boys in the hall. There were 12 more boys than girls. Then 15 more girls came into the hall. How many girls are in the hall now?

```
┌─────────────────────┐
│                     │
│                     │
└─────────────────────┘
```

What numbers do you know from the question that you can write in the boxes below? What numbers are unknown?
Can you calculate the unknown numbers? Do you have to add or subtract to calculate the unknown numbers?

| More: boys or girls? How many? | | How many more? | Then | Now |
| Fewer: boys or girls? How many? | | | | |

6 Katie had some cakes. Then she bought 6 more small cakes. After that she had 17 more cakes than Azam. Azam had 23 cakes. How many cakes did Katie have in the first place?

```
┌─────────────────────┐
│                     │
│                     │
└─────────────────────┘
```

What numbers do you know from the question that you can write in the boxes below? What numbers are unknown?
Can you calculate the unknown numbers? Do you have to add or subtract to calculate the unknown numbers?

How many in the first place?	Then	Who had more? How many?	
			How many more?
		Who had fewer? How many?	

7 The table shows the length of four bridges in Scotland.

Name of bridge	Length in metres
Tay Rail Bridge	3264
Forth Road Bridge	2512
Erskine Bridge	1321
Skye Bridge	570

a. What is the difference in length between the Tay Rail Bridge and the Erskine Bridge?

metres

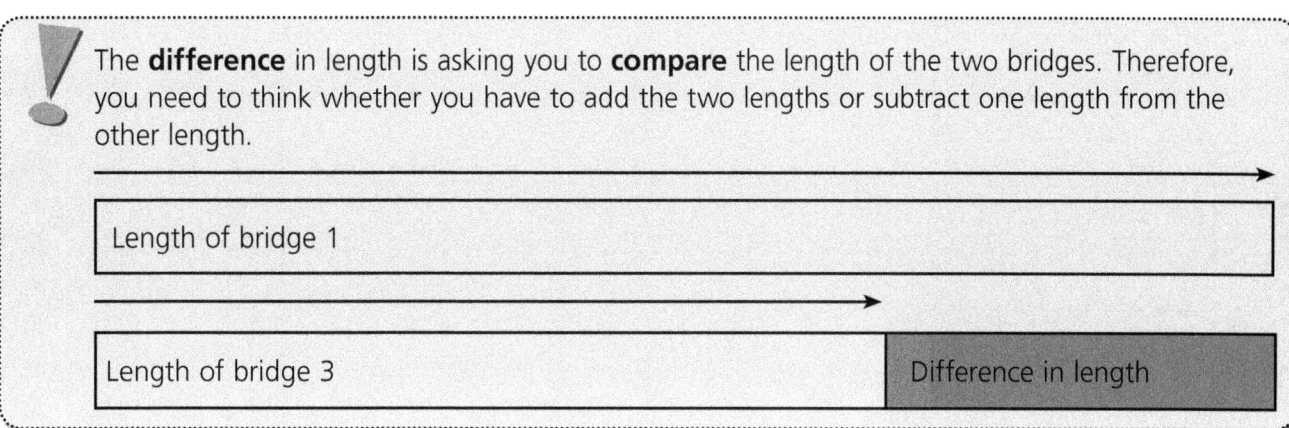

The **difference** in length is asking you to **compare** the length of the two bridges. Therefore, you need to think whether you have to add the two lengths or subtract one length from the other length.

Length of bridge 1	

Length of bridge 3	Difference in length

b. How much longer is the Tay Rail Bridge than the combined lengths of the Forth Road Bridge and the Skye Bridge?

metres

How much longer is again asking you to **compare**. This time you have to **compare** one bridge to the **combined** lengths of two bridges. **Combined** means the lengths of the two bridges added together.

Length of bridge 1		

Length of bridge 2	Length of bridge 4	Difference in length

8 I think of a number and subtract 12. My number now is 27 more than 16. What number did I think of in the first place?

> What numbers do you know from the question that you can write in the boxes below? What numbers are unknown?
>
> Can you calculate the unknown numbers? Do you have to add or subtract to calculate the unknown numbers?
>
> | What number? | → | Then | → | What number? |
>
> How many more?
>
> What number?

9 Ricardo had these four digit cards.

| 6 | 9 | 5 | 4 |

a. He made two numbers, each with two digits. He added them together. The answer was 150. What two numbers did he make?

[ ][ ]   and   [ ][ ]

> How many **ones** are there in 150? What two numbers when you add them together will give you zero **ones**?

b. He then made two different two-digit numbers. This time, he subtracted the smaller from the larger number. The answer was 16. What were the two new numbers?

[ ][ ]   and   [ ][ ]

> How many **ones** are there in 16? What two numbers when they are subtracted from each other will give you 6 **ones**?

a. The numbers in this sequence increase by 17 each time. Fill in the missing numbers.

8		42		76		110

 In this example, the numbers in the sequence **increase by 8** each time.

12	20	28	36	44	52	60

Each number is **8 more** than the number before.

Therefore 12 + 8 = 20 and then 20 + 8 = 28, 28 + 8 = 36 and so on.

b. The numbers in this sequence decrease by the same number each time. Fill in the missing numbers.

82		56		30	17	

In another example, the numbers in the sequence **decrease by 9** each time.

94	85	76	67	58	49	40

Each number is **9 less** (or fewer) than the number before.

Therefore 94 − 9 = 85 and then 85 − 9 = 76 and so on.

Answers

1. £20

2. 37

3. 75

4. 15

5. 57

6. 34

7. a. 1943 metres

   b. 182 metres

8. 55

9. a.

| 9 | 6 | + | 5 | 4 |

or

| 5 | 6 | + | 9 | 4 |

  b.

| 6 | 5 | − | 4 | 9 |

10. a.

| 8 | **25** | 42 | **59** | 76 | **93** | 110 |

   b.

| 82 | **69** | 56 | **43** | 30 | 17 | **4** |

The numbers decrease by 13 each time.

Number – multiplication and division

Collaborative activities

Information gap / barrier game: Multiplication and division

An activity for two pupils. *See download for accompanying printable resources.*

Content: Matching division and multiplication cards with the same answer

Key language

Three **multiplied by** two has the same answer as thirty-six **divided by** six.

Instructions

1. Pupil A takes the division table and the set of cut-up division cards. Pupil B takes the multiplication table and the set of cut-up multiplication cards.

2. Each pupil places cards over the squares that match on their table. Each pupil will be left with six cards and now needs to find out where they go on the table.

3. Pupil B says: "I need to know what goes in the first square in the top row."

4. Pupil A says: "36 ÷ 6. Which multiplication card gives you the same answer?"

5. Pupil B works out that 36 ÷ 6 = 6 and that 3 × 2 gives the same answer and says: "I think that 3 × 2 gives the same answer." If Pupil A agrees, the card can be placed in the empty square.

6. The roles are then reversed, with Pupil A saying: "I need to know what goes in the middle square in the top row."

7. Pupil B says: "5 × 5. Which division card gives you the same answer?"

8. Pupil B works out that 125 ÷ 5 gives the same answer (25) and says: "I think that 125 ÷ 5 gives the same answer." If Pupil A agrees, the card can be placed in the empty square.

9. Each pupil then plays in turn until both tables are completed.

Four in a line: Multiplication and division

An activity for two pupils or four pupils in two teams of two. *See download for accompanying printable resources.*

Content: Multiplication and division

Key language

Three **multiplied by** twenty-five equals seventy-five.

Two hundred **divided by** five is forty.

Instructions

1. Cut the cards into individual cards.

2. One pupil or team has the yellow cards and the other pupil or team has the green cards.

3. Place the cards face down in two piles.

4. The yellow team turns over a card and places it on the box which is the answer to the question on the card, saying aloud why the card goes on that square (e.g. "Three multiplied by twenty-five equals seventy-five.").

5. The green team turns over a card and places it on the box which is the answer to the question on the card, saying aloud why the card goes on that box (e.g. "Two hundred divided by five is forty.").

6. The aim of the game is to get four cards of your colour in a line – across, down or diagonally.

7. The winner is the first pupil or team to get four of their colour in a line.

Domino chains: Multiplication

An activity for the whole class or four pupils in two teams of two. *See download for accompanying printable resources.*

Content: Multiplication

Key language

Ninety-nine is thirty-three **multiplied by** three. What is twelve **multiplied by** six?

Seventy-two is twelve **multiplied by** six. What is thirteen **multiplied by** four?

Instructions (for whole-class activity)

1. Cut out the dominoes. Each domino has a whole number on the left (e.g. 99) and a multiplication calculation on the right (e.g. 12 × 6).

2. Give a domino to each pupil.

3. Choose a pupil to read out the calculation on their domino (e.g. "Twelve multiplied by six").

4. The pupil with the domino containing the answer on the left-hand side says: "I have the answer. It is seventy-two. The calculation on my domino is thirteen multiplied by four."

5. The pupil with the domino containing the answer to this calculation says: "I have the answer. It is fifty-two. The calculation on my domino is twenty multiplied by thirty."

6. Repeat until all the dominoes have been used.

Instructions (for four pupils)

1. Cut out the dominoes.

2. Divide the dominoes equally between the four pupils.

3. The pupil with the number 99 puts the domino down on the table and reads the calculation on the domino: "Twelve multiplied by six".

4. The pupil with the domino containing the answer (72) on the left-hand side says: "I have the answer. It is seventy-two. The calculation on my domino is thirteen multiplied by four." The pupil places the number part of their domino next to the calculation on the previous pupil's domino.

5. The pupil with the domino containing the answer to the second calculation says: "I have the answer. It is fifty-two. The calculation on my domino is twenty multiplied by thirty." The pupil places the number part of their domino next to the calculation on the previous pupil's domino.

6. Repeat until all the dominoes have been used.

Clue sheets: Multiplication and division word problems

An activity for two pupils. *See download for accompanying printable resources.*

Content: Solve multiplication and division two-part word problems

Key language

The school bought 36 **packets of** pencils. There were 9 pencils in **each packet**. What was **the total number of** pencils? **The total number of** pencils was 324.

The museum had **four times as many** visitors on Saturday.

Anya saved £8 **per week**.

How many … **each day / week? A seventh / sixth / fifth / quarter / third / of** …

Context

Pupils complete a table to show the known numbers and missing numbers in different word problems. They use the 'clues' to complete the table. The problems represent typical contexts for two-part multiplication and division word problems.

Instructions

1. Pupil A has clue sheet A and Pupil B has clue sheet B. They each have a recording sheet.
2. Pupils take turns to read aloud a word problem from their clue sheet. Pupils identify the name of the person or organisation in the word problem and which line of the recording sheet has that name.
3. Pupils decide together where the given numbers should be written on the recording sheet and both write the numbers in the correct places.
4. Pupils then decide together how to calculate the missing numbers which answer the questions in the word problem.
5. Pupils write the missing numbers in the correct spaces on the recording sheet.
6. Ask pupils to explain their decision (e.g. "because there were 3600 cakes and 8 cakes in each box").

Card sorting: Multiplication

An activity for four pupils. *See download for accompanying printable resources.*

Content: Calculations involving addition and subtraction of two multiplication calculations

Key language

7 **multiplied by** 3 **plus** 7 **multiplied by** 2 equals 35.

10 **multiplied by** 9 **minus** 2 **multiplied by** 9 equals 72.

7 multiplied by 3 plus 7 multiplied by 2 **equals / is the same as / is equivalent to** 7 multiplied by 5.

10 multiplied by 9 minus 2 multiplied by 9 **equals / is the same as / is equivalent to** 8 multiplied by 9.

Context

This is a sorting activity in which pupils have to match various combinations of additions and subtractions of two multiplication calculations to a specific number. The point here is to enable pupils to understand and express the fact that, for example, 7 multiplied by 3 plus 7 multiplied by 2 equals / is the same as / is equivalent to 7 multiplied by 5.

Instructions

Note: It is preferable to enlarge the cards and sorting boards to A3 size.

1. Cut the cards into individual cards.

2. Give Pupil A the purple cards, Pupil B the blue cards, Pupil C the green cards and Pupil D the pink cards.

3. Give the group a copy of one of the sorting boards. There is a choice of two boards. Board 1 shows where the different coloured pupil cards are arranged. Board 2 does not show this and so is more challenging.

4. The pupils take turns to place one of their cards in one of the boxes surrounding the numbers. The card must be correct for that number. The pupil has to explain why it is correct. (Pupils can use small pieces of reusable adhesive to fix the card to the board.)

   Note: For the number 35 there are two blue cards, two purple cards, one pink card and one green card. For other numbers there are different combinations of colours. The aim is to get pupils to scan their cards to find correct calculations to match the numbers.

5. Pupils continue until they have placed all their cards correctly.

6. The pupils can then work together to add some of their own examples to the board. They can write these examples on the white cards and place the white cards on the white boxes surrounding the numbers.

Number - multiplication and division

Questions with support

1 Write the missing number to make this division correct.

$$95 \div \boxed{} = 9.5$$

> **Example**
>
> $$27 \div \boxed{} = 4.5$$
>
> $$27 \div 4.5 = 6$$
>
> $$27 \div \boxed{6} = 4.5$$

2 Some brothers are given £100 by their grandmother. They share the money equally. They get £25 each. How many brothers are there?

> **Example**
>
> A group of friends earns £80 by washing cars. They share the money **equally**. They get £16 **each**. How many friends are in the group?
>
> When money is shared **equally**, **each** person gets the **same amount** of money.
>
> £80 ÷ £16 = 5 **equal** shares or £16 for **each** friend. So, there are 5 friends in the group.

3 A square number and a prime number have a total of 16. What are the two numbers?

$$\boxed{} + \boxed{} = 16$$

square number   prime number

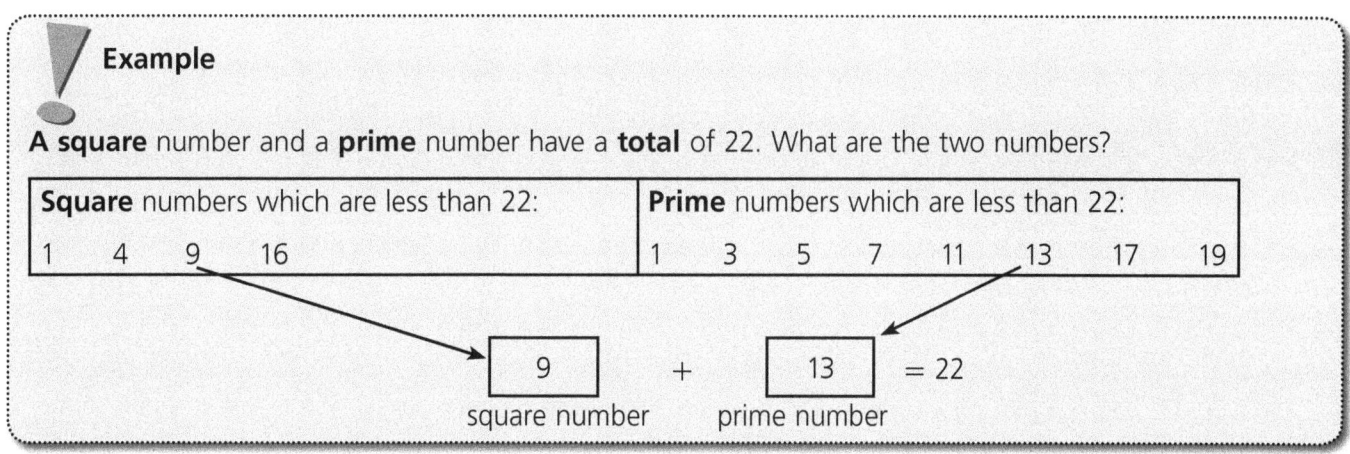

> **Example**
>
> A **square** number and a **prime** number have a **total** of 22. What are the two numbers?
>
> | **Square** numbers which are less than 22: | | | | **Prime** numbers which are less than 22: | | | | | | | |
> |---|---|---|---|---|---|---|---|---|---|---|---|
> | 1 | 4 | 9 | 16 | 1 | 3 | 5 | 7 | 11 | 13 | 17 | 19 |
>
> $$\boxed{9} + \boxed{13} = 22$$
>
> square number   prime number

4 There are 5200 leaflets in a box. Shah and Adam take 900 leaflets each. Yveta and Anna share the rest of the leaflets equally. How many leaflets does Anna get?

Show your method.

Show your method	

5 Anna uses these digit cards.

4	3	5

She makes a two-digit number and a one-digit number.

She multiplies them together. Her answer is a multiple of 10.

What could Anna's multiplication be?

Remember: **'could'** means there might be more than one correct answer.

Example

Adam uses these digit cards.

7	8	5

He makes a two-digit number and a one-digit number.

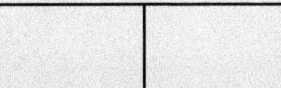 

He multiplies them together. His answer is a multiple of 10.

What could Adam's multiplication be? 7

$75 \times 8 = 600$ ✓ $85 \times 7 = 595$ $78 \times 5 = 390$ ✓

$57 \times 8 = 456$ $58 \times 7 = 406$ $87 \times 5 = 435$

$75 \times 8 = 600$ or $78 \times 5 = 390$

Answers

1. 10
2. 4
3. 9 and 7
4. 1700 ($2 \times 900 = 1800$; $5200 - 1800 = 3400$; $3400 \div 2 = 1700$)
5. 35×4 or 34×5

Number – fractions

Collaborative activities

Card sorting: Decimal sorting

An activity for two, three or four pupils. *See download for accompanying printable resources.*

Content: Decimal fractions: multiplication and division by 10, 100 and 1000

Key language

Zero point two five **divided by** ten **equals** zero point zero two five.

Zero point zero two five **multiplied by** ten **equals** zero point two five.

Instructions

1. Cut the cards into individual cards.
2. Share out the cards among the pupils.
3. Pupils take turns to read one of their cards. The pupil on the right of the reader has to say where on the sorting grid the card should be placed and why (e.g. "Zero point two five divided by ten equals zero point zero two five so it should go in the first column.").
4. The other pupils say whether they agree or not. Everybody has to agree what the answer is.
5. They place the card in the correct column on the sorting grid.
6. The next pupil reads a card.
7. Encourage pupils to justify their decisions by explaining their reasoning.

Four in a line: Decimal fractions

An activity for two pupils or four pupils in two teams of two. *See download for accompanying printable resources.*

Content: Reading and writing decimal numbers as fractions

Key language

Zero point two five is **equivalent** to a quarter.

Zero point four is **less than** a half because zero point five is **equivalent** to a half.

Zero point eight five is **more than** three-quarters because three-quarters is **equivalent** to zero point seven five.

Instructions

1. Cut the cards into individual cards. Place the cards face down in two piles.
2. One pupil or team has the yellow cards and the other pupil or team has the green cards.
3. The yellow team takes the top card from their pile and places it on a possible box on the board. When they put the card down, they must say a correct sentence (e.g. "Zero point two five is equivalent to a quarter.").
4. The green team then places their card and says a sentence.
5. The aim of the game is to get four cards of your colour in a line – across, down or diagonally.
6. The winner is the first team to get four of their colour in a line.

Information gap / barrier game: Decimals and fractions

An activity for two pupils. *See download for accompanying printable resources.*

Content: Equivalent fractions and decimals

Key language

What fraction is the **same as** this decimal?

What decimal is the **same as** this fraction?

What fraction is **equivalent** to this decimal?

What decimal is **equivalent** to this fraction?

Instructions

1. Pupil A takes the fraction table and the set of cut-up fraction cards. Pupil B takes the decimal table and the set of cut-up decimal cards.

2. Each pupil places cards over the squares that match on their table. Each pupil will be left with six cards and now needs to find out where they go on the table. This part of the activity may take some time because each pupil has to make six correct matches.

3. The pupil with the fraction table says: "I need to know what goes in the first square in the top row."

4. The pupil with the decimal table looks at their table and says: "0.5. What fraction is the same as this decimal?"

5. The pupil with the fraction table works out the fraction equivalent of 0.5 and says: "I think that is equivalent to $\frac{1}{2}$." If the pupil with the decimal table agrees, the pupil with the fraction table finds that card and places it in the empty square.

6. The roles are then reversed, with the pupil with the decimal table saying: "I need to know what goes in the middle square in the top row." The pupil with the fraction table replies: "$\frac{756}{1000}$. What decimal is equivalent to this fraction?"

7. The pupil with the decimal table works out the decimal equivalent of $\frac{756}{1000}$ and says: "I think that is equivalent to 0.756." If the pupil with the fraction table agrees, the pupil with the decimal table finds that card and places it in the empty square.

8. Each pupil then plays in turn until both tables are completed.

Domino chains: Fractions of whole numbers

An activity for the whole class or four pupils in two teams of two. *See download for accompanying printable resources.*

Content: Calculating fractions of whole numbers (e.g. $\frac{1}{3}$ of 12 is 4)

Key language

What is $\frac{1}{3}$ of 12? $\frac{1}{3}$ of 12 is 4.

Instructions (for whole-class activity)

1. Cut out the dominoes. Each domino has a whole number on the left (e.g. 5) and a fraction of a whole number on the right (e.g. $\frac{3}{4}$ of 12).

2. Give a domino to each pupil.

3. Choose a pupil to read out the first calculation on their domino (e.g. "Three-quarters of 12").

(4) The pupil who has the domino containing the answer on the left-hand side says: "I have 9. The calculation on my domino is $\frac{1}{3}$ of 3."

(5) The pupil who has the answer on their domino says: "I have 1. The calculation on my domino is $\frac{1}{3}$ of 12."

(6) Repeat until all the dominoes have been used.

Instructions (for four pupils)

(1) Cut out the dominoes.

(2) Divide the dominoes equally between the four pupils.

(3) The first pupil places a domino and says, for example: "The calculation on my domino is $\frac{1}{5}$ of 100."

(4) The pupil who has the number 20 on their domino places the number end of the domino next to the fraction end of the previous domino and says: "I have 20. The calculation on my domino is $\frac{1}{6}$ of 36."

(5) Repeat until one pupil has used all their dominoes or until all the dominoes have been used.

Four in a line: Rounding decimals

An activity for two pupils or four pupils in two teams of two. *See download for accompanying printable resources.*

Content: Rounding decimals to the nearest whole number or to one decimal place

Key language

One point three three **rounded to one decimal place** is one point three.

Two point seven seven **rounded to the nearest whole number** is three.

Instructions

(1) Cut the cards into individual cards.

(2) One pupil or team has the yellow cards and the other pupil or team has the green cards.

(3) Place the cards face down in two piles.

(4) The yellow team takes the top card from their pile and places it on a possible box on the board. When they put the card down, they must say a correct sentence (e.g. "One point three three rounded to one decimal place is one point three.").

(5) The green team then places their card and says a sentence.

(6) The aim of the game is to get four cards of your colour in a line – across, down or diagonally.

(7) The winner is the first team to get four of their colour in a line.

Card matching: Decimal places

An activity for two, three or four pupils. *See download for accompanying printable resources.*

Content: Addition and subtraction of two decimals less than 1

Key language

Zero point two two five plus **zero point four seven five** equals **zero point seven**.

Zero point eight one three minus **zero point five one three** equals **zero point three**.

Note: This activity is intended to 'talk decimals' as well as to support pupils' ability to add and subtract decimals correctly.

1. Cut the cards into individual cards.

2. Give the group a copy of the answer board and a set of the cards.

3. Pupil A takes a card and reads it aloud to the rest of the group. (Make sure that pupils read the decimals correctly, i.e. "zero point six two" not "zero point sixty-two".)

4. Pupil B says what they think the answer is. Depending on the number of pupils in the group, the other pupils say whether they agree or disagree with Pupil B's answer.

5. When the correct answer has been agreed, the card is placed on the correct space on the board. Pupils continue until all the cards have been placed.

6. Pupils can then use their own copy of the answer board to write their own examples of additions and subtractions in the spaces so that they correspond with the answers.

Substitution tables: Fractions

See download for full-size printable tables.

Content: Asking questions and making statements about fractions of quantities and addition and subtraction of fractions with the same denominator

Key language

See substitution tables.

Asking questions

Use the table to create 10 questions. Ask your friend to answer your questions. Check to see if their answers are correct.

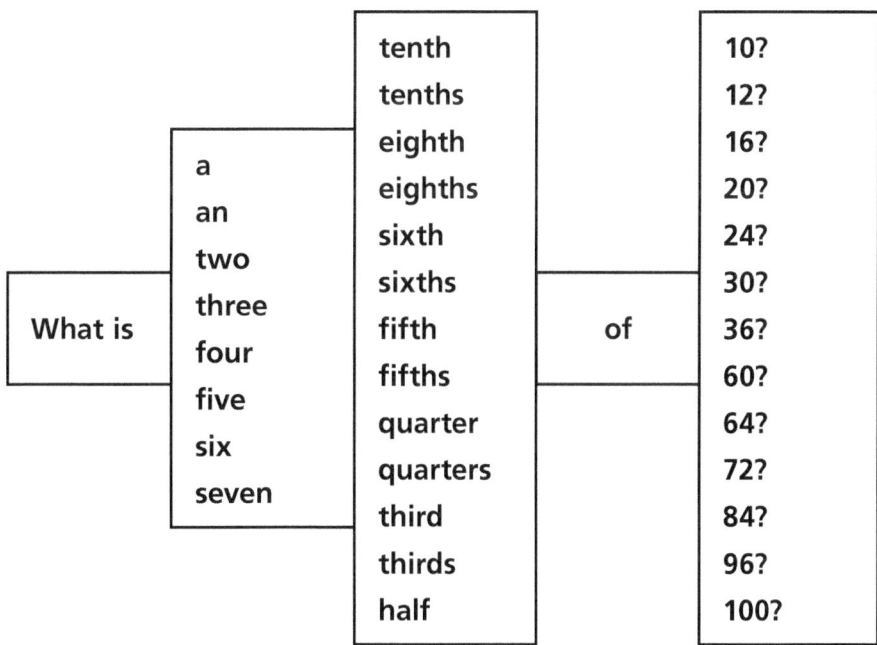

| What is | a / an / two / three / four / five / six / seven | tenth / tenths / eighth / eighths / sixth / sixths / fifth / fifths / quarter / quarters / third / thirds / half | of | 10? / 12? / 16? / 20? / 24? / 30? / 36? / 60? / 64? / 72? / 84? / 96? / 100? |

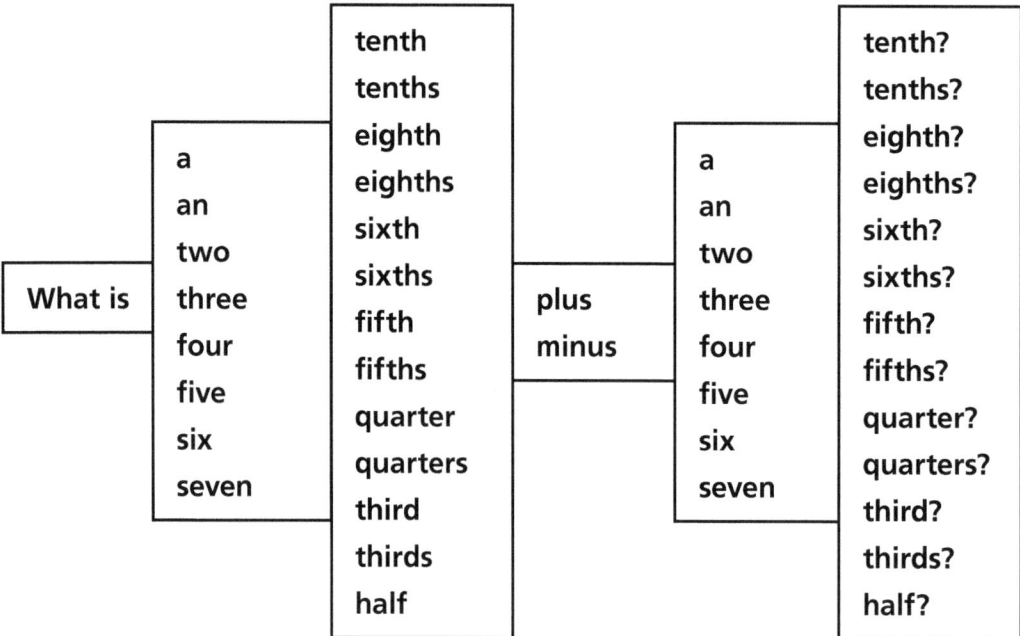

More questions

Use the table to create 10 more questions. Ask your friend to answer your questions. Check to see if their answers are correct.

Note: Ensure that pupils understand that both parts of the question must use the same denominator.

| What is | a
an
two
three
four
five
six
seven | tenth
tenths
eighth
eighths
sixth
sixths
fifth
fifths
quarter
quarters
third
thirds
half | plus
minus | a
an
two
three
four
five
six
seven | tenth?
tenths?
eighth?
eighths?
sixth?
sixths?
fifth?
fifths?
quarter?
quarters?
third?
thirds?
half? |

Number – fractions

Questions with support

1 Write the two missing values to make these equivalent fractions correct.

$$\frac{3}{\boxed{}} = \frac{6}{8} = \frac{\boxed{}}{12}$$

> ⚠ You need to remember that **values** means **numbers** so you need to write the **missing numbers** in the boxes.
>
> **Example**
>
> Write the two missing values to make these equivalent fractions correct.
>
> $$\frac{2}{\boxed{3}} = \frac{6}{9} = \frac{\boxed{8}}{12}$$

2 Yveta had some money.

She spent £2.50 on a drink. She spent £1.30 on a sandwich.

She has three-quarters of her money left.

How much money did Yveta have to start with? Show your method.

Show your method	

> ⚠ **Example**
>
> Yveta had some money. She **spent** £1.25 on a drink. She **spent** £1.60 on a sandwich. She **has three-quarters of her money left**.
>
> How much money did Yveta have to start with? Show your method.
>
> | Yveta **spent** |
> | --- |
> | £1.25 + £1.60 = £2.85 |
> | £2.85 is one-quarter ($\frac{1}{4}$) of her money because **she has three-quarters ($\frac{3}{4}$) of her money left**. |
> | So, **to start with** she had: |
> | 4 × £2.85 = £11.40 |

3 Tick (✓) two shapes that have $\frac{1}{4}$ shaded.

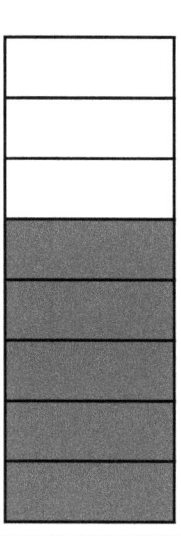

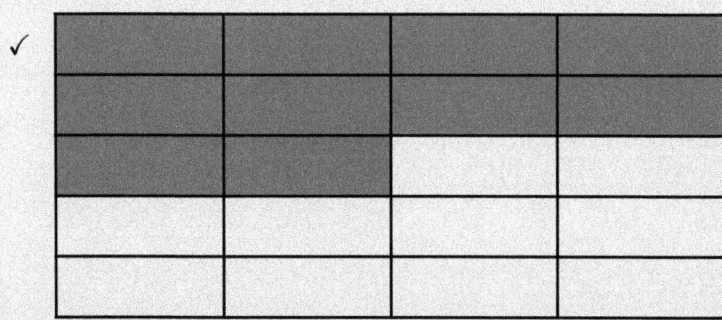
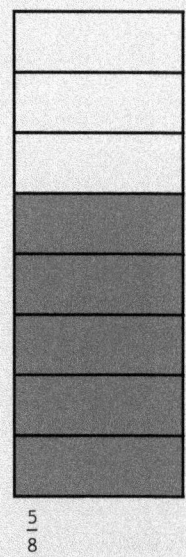

4 Circle two numbers that add together to equal 0.4.

0.35   0.7   0.29   0.05

5 Yveta chooses a number less than 30.

She divides it by 4 and then adds 6.

She then divides this result by 5.

Her answer is 2.5.

What was the number she started with? Show your method.

Show your method	

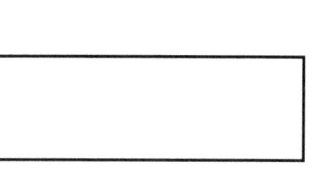

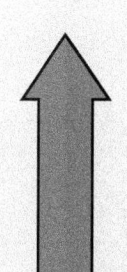

6 Write these numbers in order of size, starting with the smallest.

2.43   0.6   0.586   2.362

Smallest   **Largest**

> **Example**
>
> Write these numbers in order of size, starting with the smallest.
>
> 1.9   0.96   1.253   0.328
>
> 0.328 and 0.96 are smaller than 1.9 and 1.253 because they both start with zero (0). 0.**3**28 is smaller than 0.**9**6 and 1.**2**53 is smaller than 1.**9** so the answer is:
>
> | | | | |
> |---|---|---|---|
> | 0.328 | 0.96 | 1.253 | 1.9 |
>
> **Smallest**   **Largest**

Answers

1.

$$\frac{3}{\boxed{4}} = \frac{6}{8} = \frac{\boxed{9}}{12}$$

2. Yveta spent

£2.50 + £1.30 = £3.80

£3.80 is one-quarter $(\frac{1}{4})$ of her money because she has three-quarters $(\frac{3}{4})$ of her money left.

So, to start with she had

4 × £3.80 = £15.20

3. 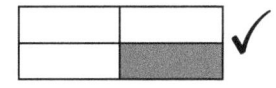 ✓ 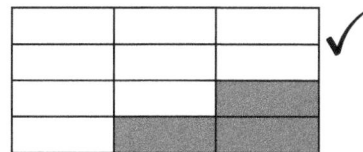 ✓

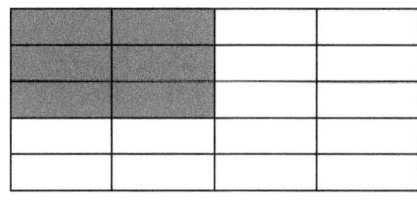 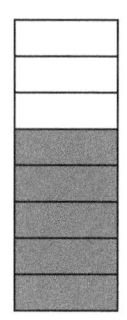

4. 0.35 and 0.05

5. 2.5 × 5 = 12.5

12.5 − 6 = 6.5

6.5 × 4 = 26

6. 0.586   0.6   2.362   2.43

Explain why

Collaborative activities

Matching activity, questions and substitution table: What coins?

An activity for two, three or four pupils. *See download for accompanying printable resources.*

Content: Expressing and explaining possibilities and logical necessities about the possible combinations of seven coins

Key language

If she has ... she could / can't / doesn't have to / must have ...

Context

Anita has seven coins. They amount to the value of £1.24. There are six different ways to make £1.24 with seven coins. Pupils use the grid to work out the six possible ways to make £1.24.

Instructions

This is intended to be a guided talk activity in which the teacher or teaching assistant works with a small group of two, three or four pupils.

1. Give each pupil a copy of the recording sheet. It is useful if pupils are provided with coins (real or plastic) to support them with the activity.

2. In pairs, pupils work out a combination of seven coins which amount to the value of £1.24.

3. They record the combination on the recording sheet.

4. They then find another combination of seven coins which amount to the value of £1.24. There are six combinations for them to find altogether.

5. When pupils have finished, encourage them to verbalise the information on the recording sheet by asking prompt questions to elicit responses. For example, pupils may be prompted to say: "She could have one pound coin, two 10p coins and four 1p coins" or "She can't have four 20p coins".

6. The pupils then use the information from the recording sheet to complete the question sheet. They then use the substitution table to create their own true statements about the information on the recording sheet.

Calculation grid and substitution table: In the supermarket

An activity for two, three or four pupils. *See download for accompanying printable resources.*

Content: Explaining possibilities and necessities about buying quantities of goods within specified budgets

Key language

She can / can't buy _ packets / tins / cartons / bags of ...

He has /doesn't have enough to buy ...

He could buy ...

She needs £_ more to buy ...

Instructions

This is intended to be a guided talk activity in which the teacher or teaching assistant works with a small group of two, three or four pupils.

1. Each pupil has a copy of the recording sheet and the price list.
2. The pupils together calculate and record the cost of the items on the people's lists. They also decide whether each person has enough money to buy all the items on their list.
3. Pupils use the substitution table to compose sentences explaining what a person can and can't buy.
4. The teacher / teaching assistant can use prompt questions (e.g. "Does he have enough money to buy ...?", "Why can't he buy ...?", "What could he buy?") to elicit explanations from pupils which use the words can / can't / could.

True or false

An activity for two, three or four pupils. *See download for accompanying printable resources.*

Content: Explanations related to various aspects of calculation, including factors, multiples, prime numbers, squared and cubed numbers, decimals and fractions

Key language

multiplied by, divided by, divisible exactly by, multiple of, factor of, prime factors of, prime number, squared, cubed, equivalent to

Instructions

This is intended to be a guided talk activity in which the teacher or teaching assistant works with a small group of two, three or four pupils.

1. Each pupil has a copy of the recording sheet.
2. Pupil A reads the first statement. Pupil B has to say whether the statement is true or false and explain why. The teacher or teaching assistant can prompt Pupil B with questions to elicit the explanation and re-model Pupil B's statements if necessary. Pupils C and D say whether they agree or disagree with the explanation. They can also repeat the explanation out loud when it is correct.
3. Pupil B then reads the next statement and Pupil C decides whether it is true or false and explains why. Pupils D and A say whether they agree or disagree.
4. Pupils write their explanations in the spaces on the recording sheet. The explanations can be scaffolded by providing mini writing frames or gap fill texts.

Note: Some of the statements are written in words (e.g. 4 squared) and some in figures (e.g. 4^2). This is to provide some variety and correspondence between reading words and mathematical notation.

Sample explanations are provided, but can be varied to suit the pupils. More formal vocabulary such as 'multiplied by' or 'divisible by' rather than 'times' or 'goes into' is used to help pupils acquire subject specific vocabulary.

Explain why

Questions with support

1. Salim bought three kinds of fruit at a shop. He bought 7 apples, 6 oranges and 8 bananas. He paid £3.04 in total for one kind of fruit, £2.46 for another and £2.52 for another. The cost of one fruit was an exact number of pence.

What was the cost of 1 orange? What was the cost of 1 banana? What was the cost of 1 apple?

> ! The cost of one fruit was an exact number of pence. If Salim bought 7 apples, could they have cost £3.04? They couldn't have cost £3.04 because £3.04 is not divisible by 7 without a remainder.
>
> Could they have cost £2.46? They couldn't have cost _____ because
>
> _____
>
> They must have cost _____ because _____
>
> Therefore, 1 apple must have cost _____.
>
> If Salim bought 8 bananas, could they have cost £2.46? They _____
>
> because_____.
>
> Therefore, they must have cost _____ because _____.
>
> Therefore, 1 banana must have cost _____.
>
> 6 oranges must have cost _____ and, therefore, 1 orange must have cost _____.

2. A shop has 324 T-shirts. The T-shirts are either red, blue or green. The shopkeeper says that 100 of them are red. He also says that he has 60 blue T-shirts. Richard says that it must mean that most of the T-shirts are red. Is he right or wrong? Can you explain why?

> ! The total number of T-shirts is _____. If _____ of the T-shirts are red, then _____ minus _____ must equal the number of blue T-shirts and green T-shirts. _____ minus _____ equals _____.
>
> Therefore, the number of blue T-shirts and green T-shirts must be _____. If there are _____ blue T-shirts, then _____ minus _____ must equal the number of green T-shirts. _____ minus _____ equals _____. Therefore, there must be _____ green T-shirts.
>
> _____ is more than 100. Therefore, Richard is _____ because most of the T-shirts are _____ and not _____.

3 There are three sisters, Amy, Beth and Caroline. Amy is 4 years older than Beth. Caroline is 7 years younger than Amy. Caroline is 10 years old. Beth is 5 years older than Diana.

Haley says that Diana must be 15. Is she right?

If Caroline is _____ years old and she is _____ years younger than Amy, then Amy must be _____ plus _____ years old. Therefore, Amy must be _____ years old. If Amy is _____ and she is _____ years older than Beth, then Beth must be _____ minus _____ years old. Therefore, Beth must be _____ years old. If Beth is _____ years old and Diana is _____ years younger than Beth, then Diana must be _____ minus _____ years old. Therefore, Diana must be _____ years old. Therefore, Haley is _____ _____ because Diana is _____ years old and not _____ years old.

4 Oliver, Patryk and Robert have 216 stickers altogether. Oliver has 90 stickers. Patryk has twice as many stickers as Robert.

Their friend Daniel says that Robert must have 48 stickers. Is he right?

Daniel is / is not right. If Oliver has _____ stickers, then Patryk and Robert must have 216 minus _____ stickers. _____ minus _____ equals _____. Patryk has twice as many stickers as Robert. If Robert has 48 stickers, then 48 multiplied by _____ equals _____. If Patryk has _____ and Robert has _____, then _____ plus _____ equals _____. This is greater than 126 and so Robert can't have _____ stickers. Robert must have 42 stickers because _____ multiplied by 2 equals _____ and 84 plus _____ equals _____.

5 There are three films at the cinema. Film A starts at 7 p.m. and lasts for 135 minutes. Film B starts at 7:15 p.m. and lasts for 130 minutes. Film C starts at 7:30 p.m. and lasts for 110 minutes.

Sofia says that Film C will be the last of the three films to finish. Is she right?

Film A starts at _____ p.m. and lasts for _____ minutes.

_____ minutes is 60 + 60 + 15 minutes. Therefore it lasts _____

hours and _____ minutes. If it lasts for _____ hours and

_____ minutes, then it will finish at _____ p.m.

Film B starts at _____ p.m. and lasts for _____ minutes.

_____ minutes is _____ + _____

+ _____ minutes. Therefore, it lasts _____ hours

and _____ minutes. If it lasts for _____ hours and

_____ minutes, then it will finish at _____ p.m.

Film C starts at _____ p.m. and lasts for _____ minutes.

_____ minutes is _____ + _____ minutes.

Therefore, it lasts for _____ hour and _____ minutes. If it lasts for

_____ hour and _____ minutes, then it will finish at

_____ p.m.

Therefore, Film _____ will be the last to finish because it finishes at

_____ p.m. and the other two films finish at _____ p.m. and

_____ p.m. Therefore, Sofia is _____.

Answers

1. The cost of one fruit was an exact number of pence. If Salim bought 7 apples, could they have cost £3.04? They couldn't have cost £3.04 because £3.04 is not divisible by 7 without a remainder.

    Could they have cost £2.46? They couldn't have cost **£2.46** because **£2.46 is not divisible exactly by 7**.

    They must have cost **£2.52** because **£2.52 is divisible exactly by 7.**

    Therefore, 1 apple must have cost **36p**.

    If Salim bought 8 bananas, could they have cost £2.46? They **couldn't have cost £2.46** because **£2.46 is not divisible exactly by 8**.

    Therefore, they must have cost **£3.04** because **£3.04 is divisible exactly by 8**.

    Therefore, 1 banana must have cost **38p**.

    6 oranges must have cost **£2.46** and, therefore, 1 orange must have cost **41p**.

2. The total number of T-shirts is **324**. If **100** of the T-shirts are red, then **324** minus **100** must equal the number of blue T-shirts and green T-shirts. **324** minus **100** equals **224**. Therefore, the number of blue T-shirts and green T-shirts must be **224**. If there are **60** blue T-shirts, then **224** minus **60** must equal the number of green T-shirts. **224** minus **60** equals **164**. Therefore, there must be **164** green T-shirts. **164** is more than 100. Therefore, Richard is **wrong** because most of the T-shirts are **green** and not **red**.

3. If Caroline is **10** years old and she is **7** years younger than Amy, then Amy must be **10** plus **7** years old. Therefore, Amy must be **17** years old. If Amy is **17** and she is **4** years older than Beth, then Beth must be **17** minus **4** years old. Therefore, Beth must be **13** years old. If Beth is **13** years old and Diana is **5** years younger than Beth, then Diana must be **13** minus **5** years old. Therefore, Diana must be **8** years old. Therefore, Haley is **not right** because Diana is **8** years old and not **15** years old.

4. Daniel is / **is not** right. If Oliver has **90** stickers, then Patryk and Robert must have 216 minus **90** stickers. **216** minus **90** equals **126**. Patryk has twice as many stickers as Robert. If Robert has 48 stickers then 48 multiplied by **2** equals **96**. If Patryk has **96** and Robert has **48**, then **96** plus **48** equals **144**. This is greater than 126 and so Robert can't have **48** stickers. Robert must have 42 stickers because **42** multiplied by 2 equals **84** and 84 plus **42** equals **126**.

5. Film A starts at **7** p.m. and lasts for **135** minutes. **135** minutes is 60 + 60 + 15 minutes. Therefore, it lasts **2** hours and **15** minutes. If it lasts for **2** hours and **15** minutes, then it will finish at **9:15** p.m.

    Film B starts at **7:15** p.m. and lasts for **130** minutes. **130** minutes is **60** + **60** + **10** minutes. Therefore, it lasts **2** hours and **10** minutes. If it lasts for **2** hours and **10** minutes, then it will finish at **9:25** p.m.

    Film C starts at **7:30** p.m. and lasts for **110** minutes. **110** minutes is **60** + **50** minutes. Therefore, it lasts for **1** hour and **50** minutes. If it lasts for **1** hour and **50** minutes, then it will finish at **9:20** p.m.

    Therefore, Film **B** will be the last to finish because it finishes at **9:25** p.m. and the other two films finish at **9:15** p.m. and **9:20** p.m. Therefore, Sofia is **not right**.

Acknowledgments

Published by Keen Kite Books
An imprint of HarperCollins*Publishers* Ltd
The News Building
1 London Bridge Street
London
SE1 9GF

Text and design © 2018 Keen Kite Books, an imprint of HarperCollins*Publishers* Ltd

10 9 8 7 6 5 4 3 2 1

ISBN 978-0-00-823854-4

Authors: Graham Smith and Steve Cooke, The EAL Academy

Series Concept and Commissioning: Shelley Teasdale and Michelle I'Anson

Project Manager/Editor: Fiona Watson

Cover Design: Anthony Godber

Text Design and Layout: QBS Learning

Production: Natalia Rebow

A CIP record of this book is available from the British Library.